my guide to the planets

Uranus

CHERRY LAKE PRESS

Published in the United States of America by Cherry Lake Publishing
Ann Arbor, Michigan
www.cherrylakepublishing.com

Reading Adviser: Marla Conn, MS, Ed, Literacy specialist, Read-Ability, Inc.
Book Designer: Jennifer Wahl
Illustrator: Jeff Bane

Photo Credits: Photo Credits: ©Vadim Sadovski/Shutterstock, 5, 13, 15, 21; ©NASA images/Shutterstock, 7; ©Yaroslav Vitovskiy/Shutterstock, 9; ©Zakharchuk/Shutterstock, 11; ©photosoft/Shutterstock, 17; ©nalyvme/Shutterstock, 19; ©NASA images/Shutterstock, 23; Cover, 2-3, 10, 12, 22, Jeff Bane; Various vector images throughout courtesy of Shutterstock.com/

Library of Congress Cataloging-in-Publication Data

Names: Devera, Czeena, author. | Bane, Jeff, 1957- illustrator. | Devera, Czeena. My guide to the planets.
Title: Uranus / by Czeena Devera ; illustrated by Jeff Bane.
Description: Ann Arbor, Michigan : Cherry Lake Publishing, [2020] | Series: My guide to the planets | Includes index. | Audience: K-1.
Identifiers: LCCN 2019032867 | ISBN 9781534158870 (hardcover) | ISBN 9781534161177 (paperback) | ISBN 9781534160026 (pdf) | ISBN 9781534162327 (ebook)
Subjects: LCSH: Uranus (Planet)--Juvenile literature.
Classification: LCC QB681 .D48 2020 | DDC 523.47--dc23
LC record available at https://lccn.loc.gov/2019032867

Printed in the United States of America
Corporate Graphics

table of contents

About the author: Czeena Devera grew up in the red-hot heat of Arizona surrounded by books. Her childhood bedroom had built-in bookshelves that were always full. She now lives in Michigan with an even bigger library of books.

About the illustrator: Jeff Bane and his two business partners own a studio along the American River in Folsom, California, home of the 1849 Gold Rush. When Jeff's not sketching or illustrating for clients, he's either swimming or kayaking in the river to relax.

About Uranus

I’m Uranus. I have 13 rings that circle me.

I am the third-largest planet. I am four times bigger than Earth.

I am the second-farthest planet from the Sun. It's pretty chilly here!

I **orbit** around the Sun. It takes me 84 years to complete 1 orbit!

I orbit differently than the other planets. I'm called the sideways planet.

I also **rotate** differently. I go in the opposite direction. Venus does this too.

I am an **ice giant**. My surface is made mostly of ice.

I am also made of gas. **Scientists** say my gas smells like rotten eggs!

I have 27 moons. Only two other planets have more moons than me. They are Jupiter and Saturn.

I am a **unique** planet. Scientists are discovering more new things about me.

glossary & index

glossary

ice giant (ISE JYE-uhnt) a giant planet composed mainly of ice

orbit (OR-bit) to travel in a curved path around something

rotate (ROH-tate) to move in a circle around a central point, like a wheel

scientists (SYE-uhn-tists) people who study nature and the world we live in

unique (yoo-NEEK) the only one of its kind

index